SAMUEL WAITE JOHNSON'S LOCOMOTIVE AESTHETIC BEAUTY - AN APPRECIATION

by Jack Braithwaite

Before discussing and illustrating the beauty and elegance of the S.W. Johnson locomotive, perhaps I may make some comments on this subject in general terms.

The British Steam Locomotive of the 1870-1914 period was universally recognised by many authorities as an example of good industrial design which combined utility with elegance of line and beauty of shape.

Even by the late 1830s and 1840s some very pretty designs had been evolved by private locomotive building firms. The 'early Sharp' and the slightly later 'little Sharp' 2-2-2 designs built by Sharp, Roberts and Company (later renamed Sharp Brothers and Company after Richard Roberts left in 1843), were highly praised. That celebrated railway historian, A. Rosling Bennett (1850-1928), recalling the South Eastern Railway engines of this type in his book *The First Railway in London*, published in 1912, said that these Sharp 2-2-2s (fig.1) were *'excessively pretty engines and when newly painted and polished they looked as if just taken out of a bandbox'* and *'if ever the South Kensington Museum get as far as a model of one of them with its tender, correct in all details, it will be easily the prettiest in the collection'*. When one considers that the museum already had models of two very beautiful designs, namely the Patrick Stirling 8ft 4-2-2 and William Adams 7ft 4-4-0 of 1890, this was praise indeed.

Fig.1: Sharp 2-2-2 of the South Eastern Railway.

In the January 1905 issue of *The Locomotive Magazine*, 'Centurion' recalled the Great Northern Railway 'Little Sharp' 2-2-2s saying *'I too, well remember the 'Little Sharps', than which, perhaps no prettier little engines were ever built'*.

H.L. Hopwood too, writing in the May 1911 issue of *The Railway Club Journal* on Sharp 2-2-2s considered that they were *'very pretty machines'*. Our beloved Midland Railway also possessed a number of these delightful and captivating locomotives (fig.2).

Fig.2: Sharp Roberts No.42 was built in November 1847 as a 2-2-2. It was re-built in July 1860 as a 2-4-0, but still retained all its Sharp features, as seen in this photograph taken at Kettering. The locomotive was renumbered 42A in December 1874, and withdrawn in July of the following year.

After Richard Roberts left the firm in 1843, Charles Beyer became head of the steam locomotive design section and would almost certainly have been responsible for the design of the 'Sharp' 2-2-2s from this time onwards. Beyer left the firm in 1854 to form his own company with Richard Peacock. Many very elegant locomotives were subsequently built by this celebrated firm, some of the most beautiful of them being exported.

It is interesting to recall that Charles Beyer, a German born engineer, was the main pioneer of creating the beauty of form of the British steam locomotive. Charles Beyer was a great engineer artist but, regrettably, is rarely given his just due today. He was the inspiration for the *'Manchester School'* of steam locomotive designs and, particularly from the 1850s to the 1880s, the various Manchester Locomotive building firms which included Sharp Stewart and Co., Beyer Peacock and Co., and W. Fairbairn and Co., built very many pretty and stylish looking engines, of which a large number went to countries overseas. Sadly, the 1840-1870 period, an age of *pretty* engines, is much neglected by historians today and also by modellers too, with but a few exceptions. Charles Beyer's aesthetic masterpiece was generally considered to be his famous 7ft 2-2-2 'D. Luiz' (fig. 3), built in 1862 for the 5ft 5¾ ins gauge South Eastern Railway of Portugal, and exhibited in London. This engine has been claimed by many to have been one of the most beautiful and elegant of all steam locomotives. Robert Weatherburn, the famous Midland Railway District Locomotive Superintendent at Kentish Town Shed, London, from

Fig.3: Charles Beyer's 2-2-2 'D. Luiz' of 1862.

May 1885 to December 1906, writing an article entitled *'The Aesthetical Development of the Locomotive'* in the December 1897 issue of the *Railway Magazine*, referred to D. Luiz as *'by far the simplest and neatest engine hitherto made, and which can with truth be held up as an aesthetic model to be studied with advantage'*. In similar vein the April 1893 issue of *Tramway and Railway World* remarked upon this locomotive's *'grace and symmetry of outline'*, and the 15th August 1924 edition of *The Locomotive Magazine* described the D. Luiz as *'a masterpiece of simplicity, pleasing proportion and directness of design'*. Many more such references were made to this engine's beauty of design. Happily, together with a lovely Charles Beyer 2-2-2 built for a Swedish railway, D. Luiz has been preserved. These engines are a fitting memorial to the genius of Charles Beyer.

Another most attractive design appeared in 1847, the famous 'Jenny Lind' **(fig.4)** built by E.B. Wilson and Co. Robert Weatherburn described them as *'trim and elegant, the 'beau ideal' of the single-wheel express'*. Colonel H.C.B. Rogers thought they *'were perhaps the prettiest engines ever to have run on British railways'*. These 2-2-2s were considered by V. Pendred, writing in *The Engineer* in 1896 as *'the handsomest locomotives that had been built up to that date'*. The sister periodical *Engineering* referred to 'Jenny Lind' as *'a very pretty-looking engine'*.

Fig.4: E.B. Wilson & Co's 'Jenny Lind' from an official Midland Railway postcard.

Samuel Waite Johnson was a pupil of James Fenton who was a partner in E.B. Wilson and Co. That very well known engineer, E.C. Poultney, writing in *The Engineer* in 1950 said of the Johnson bogie single wheelers *'the Midland single-wheelers enjoyed in their day considerable admiration, being certainly distinguished by a graceful outline peculiar to Derby practice of the period'. Johnson was (to quote from his obituary in the 19th January 1912 issue of The Engineer) "a stickler for beauty"*, a remark with which there will be general agreement. When, however, it is recalled that S.W. Johnson received his early engineering training with E.B. Wilson and Co, the creators of that masterpiece 'Jenny Lind', the symmetrical proportions of his own creations are in no way surprising'.

Continental locomotives did not come up to the aesthetically high standard achieved in Great Britain generally, although many Sharp 2-2-2s were exported to Europe, and in France and Germany private locomotive companies copied almost exactly the Sharp designs.

Mention must be made of the Crampton Locomotives built by German and French firms to the design of Thomas Crampton **(fig.5)**. As originally built these engines were highly picturesque and had a low sleek elegant outline. W.J. Bell, the editor of *The Locomotive*, wrote in the 15th January 1924 issue on the four Cramptons built by Maffei in 1853 for the Palatinate Railway in South Germany *'that for symmetry of design none has excelled the little engine illustrated'*.

Fig.5: Crampton 'Trifels' built by E. Kesssler in 1855 for the Bavarian Imperial Railway in Germany, seen here at Newstadt.

An aesthetic tradition was developed in the 19th century in the USA and such designers as William Mason, James Millholland, Thomas Ely and William Buchanan and others, built some beautiful locomotives. William Mason stood out alone and apart from them all, and from 1853 to circa 1880 a whole series of elegant engines were produced **(fig.6)**. M.N. Forney said of Mason *'he was a wonderfully ingenious man and combined with his ingenuity a high order of the artistic sense, so that his work was always most exquisitely designed. It might be said of his locomotives that they are melodies cast and wrought in metal'*.

Fig.6: A William Mason 4-4-0 from the USA. This illustration has been taken from a newspaper.

Finally, before turning towards S.W. Johnson, I would like to mention the later engines of Joseph Beattie of the London and South Western Railway (**fig.7**). William Briggs Thompson (1867-1962) writing on Beattie 2-4-0s of the later types in *The Locomotive*, for 15th April 1933, remarked that *'with their graceful outline, their bright paint and their profusion of polished brass and copper fittings they were some of the handsomest engines I have ever seen'*. In the 13th July 1933 issue, C. Chambers, the author of the classic history of London and South Western Locomotives, which had appeared in pre-World War One volumes, said *'I am also in agreement with Mr Thompson that some of the early South Western engines were exceedingly handsome'*.

Fig.7: Beattie 6ft 6in 2-4-0 'St George' of the London & South Western Railway.

From 1870 until 1900 was the age of the great artist engineers, S.W. Johnson, W. Stroudley, P. Stirling and very many others. I consider that S.W. Johnson was easily the greatest artist of the steam age. From 1866 to 1901, he created a whole series of exquisitely lovely small-boilered designs. Viewed objectively they had a faultless beauty, both in detail and overall design. Even his large-boilered Belpaires and Compounds which did not possess the sheer elegance of their smaller forebears, nevertheless bore comparison with other companies locomotives of similar size. The large-boilered 0-6-0s built from 1903 onwards were perhaps the least satisfactory of Johnson's designs aesthetically. Although neat and sturdy, there seemed to be a North Eastern Railway influence, notably in the shape of their chimney and safety valve casings. Perhaps J.W. Smith was responsible for much of the detail work of this group of locomotives. They certainly lacked the delicate winsome charm of the S.W. Johnson 0-6-0s built from 1875 to 1902.

From an artistic point of view, I think that Charles Beyer's designs greatly influenced Samuel Johnson in much the same way that, in later years, Johnson's locomotives did for John G. Robinson. Robert Weatherburn in his book *Ajax Loquitur*, published in 1899, says *'remember, as the hat is to a well-dressed man, so is the chimney to the locomotive'*. This remark has been attributed to John G. Robinson who became Locomotive Engineer for the Great Central Railway in June 1900, but I have found no evidence of this to date.

S.W. Johnson's Great Eastern locomotives were a delightful group. His small 0-4-2 tanks (**fig.8**) were particularly admired by Iain Rice, the well known modeller, who said he had long regarded them as *'one of the prettiest engines of all time'*, and he also thought S.W. Johnson *'to be an artist if ever there was one'*.

Rice also referred to the Johnson 0-4-2 tank as *'this splendidly curvaceous dame'*, and Ronald Clark also admired the *'good looks'* of these attractive engines.

Fig.8: Johnson Great Eastern Railway 0-4-2 tank No.82 in yellow livery.

The Johnson Great Eastern Railway 0-4-4 tanks (**fig.9**) were considered by O.S. Nock to have been *'a very handsome 0-4-4 suburban Tank Engine'*, and the September 1902 issue of *Locomotives and Railways* commented that *'as delivered they were painted a very pretty dark green colour with white lining and altogether had the usual smart appearance peculiar to their designer'*.

Fig.9: Johnson Great Eastern Railway 0-4-4 tank No.168.

In my opinion, the S.W. Johnson No 1 class 2-4-0s (**fig.10**) were some of the most charming of all small tender locomotives of that wheel arrangement. Certainly, they were one of Johnson's daintiest designs, possessing an inimitable and petite charm. E.L. Ahrons said that *'they were all supplied with four wheeled tenders of a very neat design and it may be said that they were the prettiest little engines that ever ran on the line. Moreover, they were as good in performance as they looked in appearance and did a wonderful amount of hard work for some forty to forty six years'*. Our former President, David F. Tee, in a letter to me, dated 13th August 1976, described them as *'most charming little engines'*.

Fig.10: Johnson No.1 Class 2-4-0 No.160 of the Great Eastern Railway.

Johnson built two series of delightful 0-6-0 tender engines for the Great Eastern Railway (figs.11 and 12), the earlier built, larger wheeled version, first produced in 1867, being really lovely little engines, full of charm and character.

Fig.11: Johnson's Great Eastern Railway 1st series 0-6-0 No.446.

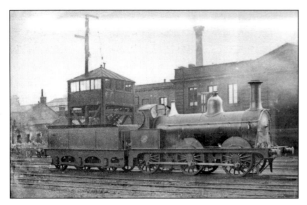

Fig.12: Great Eastern Railway 2nd design 0-6-0 No.487.

My own favourites of his Great Eastern designs were his two re-builds of Robert Sinclair 2-2-2s as 4-2-2s, the famous yellow painted 'Butterflies' (fig.13). These were small engines (only thirty three tons or so in engine only weight) but very graceful and extremely elegant, possessing a most individual outline. Three of Sinclair's 2-2-2s were also rebuilt by Johnson as 2-2-2s, again of most attractive appearance (fig.14). The two Great Eastern Johnson 4-4-0s (fig.15) were, of course, the precursors of one of the most beautiful families of all steam locomotives, the superb series of 265 small-boilered 4-4-0s built for the Midland Railway from 1876 to 1901. These two Great Eastern Railway engines were extremely stately. C. Hamilton Ellis thought *'they were lovely engines in themselves'* and, on another occasion he described them as *'very handsome engines'*.

Fig.13: No.51 – one of the famous Johnson 'Butterflies', rebuilt from a Robert Sinclair 2-2-2.

Fig.14: Another Johnson rebuild of a Sinclair 2-2-2, this time with the same wheel arrangement.

Fig.15: Johnson Great Eastern Railway 4-4-0 No.301.

For the Midland Railway, Samuel Waite Johnson, over the years from 1876 to 1881, designed a series of 135 2-4-0 express locomotives, ten with 6ft 2ins driving wheels (fig.16), forty with 6ft 6ins driving wheels (fig.17), sixty five with 6ft 9ins and twenty with 7ft driving wheels. There were subtle differences between the various batches of each of the four varieties. All possessed an extraordinary beauty of shape and grace of line and they were ideally proportioned. As the late O.S. Nock rightly observed they had *'perfect symmetry'*. Nock described No.1402 of the '1400' Class 6ft 9ins 2-4-0s (fig.18) as *'having a beauty of 'line' that has never been surpassed. Every detail was perfectly proportioned and one can see this in the handsomely shaped chimney, the curves of the dome and of the brass safety valve cover'*.

Fig.16: No.1 Class 2-4-0 No.74, with 6ft 2in driving wheels.

Fig.17: Johnson 6ft 6in 2-4-0 No.1305.

Fig.18: 1400 Class 6ft 9in 2-4-0 No.1405.

The late Brian Haresnape also considered both the Johnson 2-4-0s and small-boilered 4-4-0s **(fig.19)** to be *'a work of art'*. The late George Dow referred to S.W. Johnson and *'the aesthetic harmony of his locomotive designs'* also commenting on *'the beauty of the Johnson 2-4-0s; I don't think any other 2-4-0 could touch them'*.

Fig.19: 1400 Class 6ft 9in 2-4-0 No.1405.

The two hundred and sixty five Johnson 4-4-0s, built between 1876 and 1901 **(fig.20)** were, collectively, surely the most elegant and beautiful of this wheel arrangement ever to run on rails. The 6ft 6ins and 7ft engines were developed and gradually increased in size in this most graceful style, but the thirty 6ft 9ins engines, considered by some enthusiasts as the loveliest group of all these classically beautiful locomotives, were not perpetuated. The next 6ft 9ins 4-4-0s were Johnson's Belpaires introduced in 1900, which had a completely different outline.

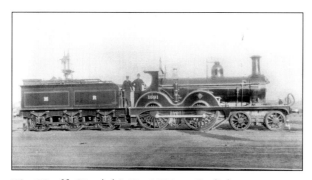

Fig.20: 6ft 9in 4-4-0 No.1581 at Carlisle.

O.S.Nock wrote *'but some very beautiful examples of the inside cylinder 4-4-0 were introduced by S.W. Johnson on the Midland Railway, which for elegance of line as well as efficiency in performance were probably unequalled anywhere in the world at that time'*.

Nock was educated at Giggleswick School during and just after World War I, and recalled his memories of Johnson's 1327 class 7ft 4-4-0s years later in the 1956 volume of Railways, saying *'what supremely beautiful engines they were!'* Towards the end of his life Nock again expressed his admiration for the 1327 class 4-4-0s **(fig.21)**, commenting that *'they were most beautiful engines to behold'*.

Fig.21: 1327 Class 7ft 4-4-0 No.1338 at Leeds.

The whole family of Johnson's small-boilered 4-4-0s possessed a wonderful combination of classic flowing curves, matchless grace, and perfect harmony of line **(fig.22)**.

Fig.22: 1668 Class 7ft 4-4-0 No.1673 at Manchester.

The late Peter Winding remarked upon *'Johnson's exquisitely proportioned early 4-4-0s'*. He was referring to the 1327 class, but the 1738 class also had many ardent admirers. Number 1757, 'Beatrice' has been thought by some to be the loveliest steam locomotive ever built. Several members of our Society are known by me to hold this viewpoint!

For myself, the 7ft 2183 class **(fig.23)** had a breathtaking beauty which I personally do not think has ever been excelled. Their soaring light matchless elegance made me refer to them, in an article I wrote for the Great Central Railway Society journal *Forward*, as the Fred Astaires of the steam locomotive world. As a former professional dance teacher I cannot think of a finer tribute to the 2183 class. The 2203 class 6ft 6ins 4-4-0s **(fig.24)** were also great beauties. That fine writer, Brian Radford, described them as *'superbly elegant'*.

W.H. Jameson, writing in 1950, recalled the 1890's, saying, *'it was always a thrill at Blackburn to see the Midland Johnson 4-4-0s in their crimson lake paint,*

Fig.23: 2183 Class 7ft 4-4-0 No.2200 taken at Bedford shed.

Fig.24: 2203 Class 6ft 6in 4-4-0 No.233 at Leeds Holbeck.

Fig.25: 1312 Class 6ft 6in 4-4-0 No.1316.

Fig.26: 1808 Class 6ft 6in 4-4-0 No.1819.

cleaned up and tallowed to a high degree, and all steel work polished bright. These old locomotives of Johnson's always looked like hounds on the leash waiting to run a race – graceful in all their lines'. He thought the Johnson 4-4-0s first of all *'for flowing lines and general handsome appearance'*. Jameson also said of the Johnson 2-4-0s and 4-4-0s and Robert Billinton's B2 class 4-4-0s of the London Brighton & South Coast Railway that they *'had really wasp like waisted smoke-boxes and these were flush with the boiler-barrel – a feature which gave all these locomotives a fine clean appearance at the front end'*. The engines Jameson saw would probably have been the 1312 class 6ft 6ins 4-4-0s **(fig.25)** and the 1808 class 6ft 6ins 4-4-0s **(fig.26)**.

The later 150 **(fig.27)** and 60 class 4-4-0s had the added attributes of an imposing, majestic stateliness, particularly the 60 class, which William Briggs Thompson considered were the most beautiful engines he had ever seen. In my opinion these locomotives were the most noble and magnificent of all the classic Johnson 4-4-0 family. J.M. Bentley has said of the 60 class 7ft 4-4-0 **(fig.28)** *'on this occasion we see evidence of S.W. Johnson's ability to design the almost perfect looking locomotives'*. Praise indeed!

That well known authority F.W. Brewer, writing in the March 1899 issue of *The Locomotive Engineer and Firemen's Journal* on the small-boilered Johnson 4-4-0 classes said *'nor have the general design and appearance of these bogie classes aroused less interest than their working'* and also *'with the proportions which Mr Johnson has given his latest 7ft engines – Number 60 class – they should prove to be of a very powerful description, in addition to being as they unquestionably are, characteristically handsome looking machines'*. P. Marshall, the editor of *The Model Engineer* saw No.2591 of the 'T' series of the 60 class **(fig.29)** on exhibition in 1901 in Glasgow, and described her as *'a splendid four coupled bogie express locomotive'*. David Tee wrote to me saying *'apropos your remarks on the 60 class (which included W.B. Thompson's tribute), the coloured drawing (Neilson Reid and Company) is a thing of beauty in itself, serving to enhance, if that were possible, the fine proportions of the engine'*. What a pity these truly wonderful-looking locomotives had such short lives in their original stately condition. Majestic beauties, these were the last of the classic small-boilered 4-4-0s. The era of the incomparably elegant Johnson small-boilered coupled express locomotive was over.

Many railway enthusiasts have claimed that S.W. Johnson's series of classic bogie single-wheeled locomotives (fig.30) were the most beautiful and elegant locomotives ever built. Their pure and uncluttered beauty of outline has captivated many locomotive students and modellers over the years.

O.S. Nock thought the whole series *'were among the most beautifully proportioned passenger locomotives ever built'*. The late Ronald Jarvis, creator of the rebuilt Bulleid pacifics, has written *'I well remember as a boy at Harpenden seeing these magnificent machines coming through at high speed, often piloting a 4-4-0 locomotive'* and also *'number 118 is one of the famous Johnson 'Singles' and is, I feel, a thing of beauty by any standard'*.

The late Ian Beattie considered the Johnson 4-2-2 as the *'Prime candidate for the coveted soubriquet "Most Beautiful Steam Locomotive Ever Built"'*. In the 1980s, the *Sunday Express* featured the fifty most beautiful objects ever made in Great Britain. This included one steam locomotive, selected by the late Dr John Coiley as *'The Most Beautiful Locomotive'*, the Johnson 4-2-2 now preserved as number 673.

Samuel Waite Johnson's Locomotive Aesthetic Beauty

Fig.27: 150 Class 7ft 4-4-0 No.207, built in 1897.

Fig.28: 60 Class 7ft 4-4-0 No.93 at Cheltenham.

Fig.29: 60 Class 7ft 4-4-0 No.806 of the 'T' series at Carlisle.

Fig.30: 7ft 4in 4-2-2 No.29.

The late Kenneth Leech has said *'all Johnson's engines were most artistic in their appearance'*, and of the bogie singles *'the 115 series were "the perfect flower" of the Johnson singles'*.

The Engineer dated 20th September 1889, published an engraving of the Johnson 7ft 6ins single-wheeler number 1853 (**fig.31**) which won the Grand Prize at the Exposition held in Paris that year saying *'a glance at our engraving is quite sufficient to show that Mr Johnson has designed an exceedingly handsome engine; and the finish and workmanship generally are of the highest class. The engine is painted the beautiful dark red which Mr Johnson has introduced with so much benefit to the pockets of the shareholders; and the large brass driving boxes have an admirable effect'*.

Fig.31: No.1853 – the winner of the Grand prize at the Paris Exposition of 1889.

W.M. (later Sir William) Acworth, comparing British (**fig.32**) and American locomotives at the Paris Exposition in *The Engineer* for 23rd December 1892, commented *'he (William Briggs Thompson) speaks of the "universal" admiration expressed for the British Exhibits at Paris in 1889 and I myself well remember the pride with which I took a party of French friends to see Mr Johnson's Midland "Single", to my mind the most beautiful type of engine I ever saw, and the satisfaction with which I contrasted it's appearance with that of an adjacent engine of the Orleans Company, which an American characterised with more accuracy than servility as "just like a darned great brewery". No doubt the foreign admiration was both genuine and deserved, but I recall another occasion when "all the world wondered" at an English Exhibition – it was the "charge of the Six Hundred"; and I am not wholly satisfied that the driver of a clumsy looking cheaply built and badly groomed American engine might not apply to Mr Johnson's beautiful exhibit the famous words "Magnificent but not War"'*.

Fig.32: 7ft 6in 4-2-2 No.179 of the third variety of five classes seen here at Worcester on a northbound express on 23rd April 1895.

That most majestic and imposing 7ft 9½ins single-wheeler number 2601 Princess of Wales won the Grand Prize at the 1900 Paris Exhibition (**fig.33**). Famous authority Charles Rous-Marten described her in *The*

One of the finest artistic masterpieces of the steam age was undoubtedly S.W. Johnson's truly wonderful bogie single-wheeler design. No.4, one of the 7ft 6in engines, shows just why so many have considered them the most elegant steam locomotives ever built.

Engineer dated 14th September 1900, saying she *'looks what she is – a railway greyhound. Her singular gracefulness and beauty of form, her relatively slender boiler-barrel, her large driving wheels 7ft 9½ins in diameter, all tend to emphasise the "greyhound" idea, and the considerable tractive force which she possesses in addition to her speed capabilities is to a great extent rendered latent by the peculiar symmetry of her design'*. Also reference was made to *'the superb and even delicate finish which characterises these beautiful engines, even in the minutest detail'*.

Fig.33: 'Princess of Wales' Class 4-2-2 No.2603.

That outstanding railway historian, the late Professor Jack Simmons, considered that the glorious family of Johnson bogie single wheelers **(fig.34)** *'are among the glories of the Victorian age'*, and he also thought *'they were surely among the most elegant locomotives ever built'*. W.B. Thompson recalling the Johnson singles said *'they were undoubtedly very beautiful engines'*.

Fig.34: A lovely 7ft 9in single-wheeler – 115 Class 4-2-2 No.128, seen here at Bristol.

Samuel Johnson's superbly proportioned 0-4-4 tank engines, built for the Midland **(fig.35)**, and Somerset & Dorset Joint **(Fig.36)** Railways have been admired by many.

John Minnis has rightly considered the Midland Railway 0-4-4 tank **(fig.37)** as *'a thing of great beauty'*. John Rowland refers to *'Johnson's delightful 0-4-4 tanks – one can say of steam locomotives as Canon Doyle did for the ladies "there are no ugly ones, some are just more beautiful than others!"'*. The late Norman Harvey described them, in one of his many articles, as *'these most attractive engines'*. Dr Charles Fryer too thought the Johnson 0-4-4 tanks to be *'very graceful tank engines'*.

Fig.35: Johnson 0-4-4 tank No. 1723.

Fig.36: Somerset & Dorset Joint 0-4-4 tank No.32.

Fig.37: Johnson 0-4-4 tank No. 1280.

The small-boilered Johnson 0-6-0 tender engines built for the Midland **(fig.38)**, Somerset & Dorset Joint **(fig.39)**, and the Midland & Great Northern Joint **(fig.40)** Railways, in my opinion, had an elegance and beauty of line and shape no others approached. Their pure chaste lines have been much admired, but I have dealt at length on these engines, and the delectable 4ft 6ins, smaller engines **(fig.41)** – these latter ones were only built for the Somerset & Dorset Joint Railway – in the *Midland Railway Society Journal No12*.

The Johnson 0-6-0 tank locomotives had the same purity of line and again were most beautifully proportioned. They were gradually increased in size, but for me the first series, the 1102s built from 1874 to 1876 **(fig.42)**, were particularly elegant and most attractive little engines. Johnson's superb artistry applied to all designs **(figs.43 & 44)**, including rebuilds of other designers work.

Fig.38: Johnson Midland 0-6-0 No.1440.

Fig.39: 5ft 2in 0-6-0 No.62 of the Somerset & Dorset Joint Railway at Bournemouth West in August 1900.

Fig.40: M&GN Joint 0-6-0 No.65.

Fig.41: S&D joint 4ft 6in 0-6-0 No.36.

Fig.42: 1102 Class 0-6-0 tank No.2255.

Fig.43: 0-6-0 tank No.2456.

Fig.44: 0-6-0 tank No.2253.

Mention must be made too, of the forty most beautiful Johnson 6ft 6ins 4-4-0 express locomotives built for the Midland & Great Northern Joint Railway (fig.45). Their classic beauty, in my opinion, made them the most elegant passenger engines ever to run on the 'Joint'. Derek Middleton has said that *'these locomotives always looked very beautiful'*.

On the Somerset & Dorset Joint Railway, eight exquisitely dainty 5ft 9ins 4-4-0s were built from 1891 to 1896 (fig.46). These were the prettiest of the small-boilered Johnson 4-4-0s, and certainly the daintiest. I just find them completely enchanting. Three large-boilered 4-4-0s followed in 1903, but these lacked the loveliness of shape and form that so distinguished the 5ft 9ins engines.

To some railway enthusiasts, Johnson's rebuilt Kirtley engines (fig.47) had even more appeal than Johnson's original superbly proportioned designs. The highly picturesque Kirtley features blended so harmoniously with the sheer elegance of Johnson's style, to make many of these Kirtley rebuilds completely irresistible to the eyes of their beholders. The last Kirtley 2-4-0s, the 890 class, of course when rebuilt presented a much different appearance to the other Johnson Kirtley rebuilds. To all intents and purposes the rebuilt 890s were perfectlooking Johnson 2-4-0s except for very minor detail differences.

The rebuilt Kirtley 2-2-2s were all most attractive little engines, but one stands out above all the others, your president's favourite railway engine, this was number 33, for so many years the Directors' saloon engine (fig.48). This exquisitely beautiful little engine was, in my opinion the prettiest small steam locomotive ever built. As that very fine modeller, Stephen Ross has written *'what a delightful locomotive number 33 is! A better example of balance proportion and symmetry would indeed be difficult to find. What perfection indeed!!'* Our former Chairman, David Hunt, holds a similar viewpoint on this enchantingly lovely steam locomotive.

Fig.45: M&GN Joint 6ft 6in 4-4-0 No.47.

Fig.46: S&D Joint 5ft 9in 4-4-0 No.17.

Fig.47: Rebuilt Kirtley 890 Class 2-4-0 No. 898.

Fig.48: Rebuilt Kirtley 2-2-2 No.33, with the Directors' saloon at Morecambe.

The rebuilt Kirtley 2-4-0s of the 50, 230 and 170 classes (figs.49, 50 & 51) had a delightful appearance, but it was the 156 (fig.52) and 800 (fig.53) classes (including the six engines of the slightly smaller 60 class) which perhaps, had the greatest number of admirers. C. Hamilton Ellis described the 156s and 800s as *'very pretty engines to look upon'* and also remarked that *'another extraordinary thing was the elegance that distinguished the old Kirtley engines after Johnson had rebuilt them, especially the 800 class expresses'*. O.S. Nock has, with truth, written that Johnson turned them *'from functional masterpieces into works of art'*. That very well known writer Eddie Johnson has said of the rebuilt 800 class 2-4-0s that *'the sheer finesse and elegance of this class is quite overwhelming – poise and charm come over in equal measure – a magnificent looking locomotive'*. I too share all the sentiments expressed above.

The Kirtley 1070 class 2-4-0s (fig.54), which appeared after Kirtley's death, when rebuilt by Johnson (like the rebuilt 890 class) presented an appearance of a typically beautiful and elegant Johnson 2-4-0. The rebuilt 890 and 1070 classes thus did not show the perfect blending of Kirtley and Johnson styling which the 156 and 800 classes so completely exemplified. The rebuilt Kirtley 0-6-0s, both straight (fig.55) and curved (fig.56) framed versions did of course, and were truly lovely engines with a charm all of their own, certainly to my eye, the most picturesque 0-6-0 tender engines ever built. David Hunt always says if he were allowed one minute in the past the one engine he would like to see is a rebuilt Kirtley 0-6-0 with the beautiful Johnson chimney and boiler fittings, in the fully lined rich red Midland livery. The rebuilt Kirtley 0-4-4 back tanks (fig.57) are also in the front rank of picturesque-looking locomotives.

Yet another lovely design were the Johnson rebuilt Charles Beyer 0-6-0 tanks. That lover of beautiful locomotives, Howard Turner, has described a picture of number 885A (fig.58) in the following terms *'what an exquisite sight number 885A makes as she stands at Kentish Town resplendent in a fresh coat of paint with lining out that can only be described as refined excellence!'*, and to Johnson's attitude *'that whatever the locomotive, it should look absolutely right and also be a first class advert for the Railway Company'* and on these class 880 0-6-0 tank rebuilds as *'some of the most attractive little shunting locomotives'*.

Fig.49: Rebuilt Kirtley 50 Class 2-4-0 No.86A.

Fig.50: Rebuilt Kirtley 230 Class 2-4-0 No.235A.

Fig.51: Rebuilt Kirtley 170 Class 2-4-0 No.181.

Fig.52: Rebuilt Kirtley 156 Class 2-4-0 No.161A at Castle Ashby L&NW.

Fig.53: Rebuilt Kirtley 800 Class 2-4-0 No.815A.

Fig.54: Rebuilt Kirtley 1070 Class 2-4-0 No.1070 at Peterborough.

Fig.55: Straight framed Kirtley 0-6-0 No.338 as rebuilt by Johnson.

Fig.56: Rebuilt Kirtley curved frame 0-6-0 No.323.

Fig.57: Rebuilt Kirtley 0-4-4 back tank No.779.

Fig.58: Rebuilt Beyer 0-6-0 tank No.885A

On the Somerset & Dorset Joint Railway, Johnson rebuilt six John Fowler designed 0-6-0s into very charming little engines (fig.59). These all retained the 'Stirling' style of cab, and number 19, the first to be rebuilt, was unique in having a polished brass dome casing, which combined with her Johnson chimney and safety valve casing made for a very pretty engine. Two James Cudworth double-framed 2-4-0s were also rebuilt with Johnson chimneys and boiler mountings (fig.60). Again, a very pretty appearance was achieved.

Fig.59: John Fowler S&D Joint 0-6-0 No.20 as rebuilt by Johnson.

Fig.60: S&D Joint line rebuilt James Cudworth 2-4-0 No.17A

Fig.61: Ex-Severn & Wye 0-6-0 tank No.1126A at Cheltenham in its rebuilt form.

The Kirtley 0-6-0 tanks rebuilt by S.W. Johnson also created a series of absolutely fascinating locomotives with many subtle differences between them. Some Severn & Wye engines were rebuilt by Johnson, number 1126A, an 0-6-0 tank being a particularly lovely example (fig.61). Johnson's 0-4-0 tanks, with their stove-pipe chimneys, also had a quaint charm.

Johnson's Belpaire (fig.62) and compound (fig.63) 4-4-0s possessed a splendid modern appearance and were locomotives of fine proportions, and although the incomparable grace and elegance of Johnson's small-boilered designs had vanished, they still had a beautiful clean-lined appearance and a distinctly Edwardian air about them.

Fig.62: Johnson Belpaire 4-4-0 No.2608.

Fig.63: Johnson Compound 4-4-0 No.2634.

To conclude, I do hope the photographs illustrated in this publication will recapture for us all the sheer glamour of appearance of the locomotives of Samuel Waite Johnson. To my mind he was the finest artist of the steam age.

BRIEF DETAILS OF PERSONS QUOTED

Sir William Mitchell Acworth: 1850-1925. Even today, he is acknowledged as one of the greatest writers on all aspects of railways and was always noted for his beautiful, clear, English.

Ernest Leopold Ahrons: 1866-1926. One of the world's greatest authorities on the steam locomotive.

Fred Astaire: 10th May 1899-1987. Considered by many to be the finest dancer of the twentieth century who acknowledged that the steam locomotive from his childhood made a lasting impression on him. Night and day, he and his sister, Adele, felt the powerful and regular beat of the mighty locomotives rumbling from coast to coast (at Omaha), each train creating its own syncopation as it gathered speed over the trackplates with punctuating, ever- altering bursts of hissing steam. They heard the irregular and incessant clanging of engines and trucks being shunted and the noise of the vast couplings clattering together. In the daytime Adele and Freddie would play at being trains, tapping their feet to the ch-ch-ch-ch-choo-choo sounds; at night the background battery would lull them into a deep and untroubled sleep. Whilst touring throughout the United States, travelling from venue to venue by train, only fifteen years of age with his seventeen years old sister, Adele, he would chat with the porters and the engineers, who were always happy to welcome the boy to their cabs. Astaire has acknowledged that the only strong masculine influences of his youth came from the train drivers on these travels. (His father also being absent from the family home through work).

Ian Beattie: Died suddenly 2000. For many years he contributed to the *Railway Modeller*. His delightful articles and drawings were admired by many, including the writer.

W.J. Bell: T the famous editor of *The Locomotive* magazine, one of the three famous Bell brothers who founded the Locomotive Publishing Company.

Rosling Bennett: 1850 –1928. A leading authority on early steam locomotives from 1840 to 1875.

J.M. Bentley: An author and former locomotive driver with a great knowl edge of the London & North Western Railway.

F.W. Brewer: A very well known writer on railways, his articles appearing in periodicals from the late nineteenth century until the late nineteen thirties.

'Centurion': A contributor who could recall the early days of the rail ways of this country from the 1840's onwards.

C. Chambers: An authority on the London & South Western Railway and the author of the superb history of the locomotives of that railway which appeared in pre-World War One volumes of *The Locomotive* magazine.

Ronald Clark: A former President of The Newcomen Society and the Midland and Great Northern Joint Railway Circle.

Dr John Coiley: The first curator of the National Railway Museum at York, also a well known writer on railways and locomotives.

George Dow: A Fellow of the Royal Society of Arts. A distinguished historian and railway officer.

Canon Doyle: Mentioned in letter by John Rowland – see below.

Cuthbert Hamilton Ellis: 1909-1987. A leading railway and locomotive historian and artist.

M. N. Forney: Celebrated American locomotive engineer who designed 0-4-4 tank engines for the elevated railways of New York.

Dr. Charles Fryer: A well known modern writer responsible for several books on steam locomotives.

Brian Haresnape: A very well known artist, aesthete and author who had a great interest in steam locomotive aesthetics.

Norman Harvey: The well known contributor during the nineteen sixties to magazines on the steam locomotive.

H.L. Hopwood: A professional railwayman who worked for the GNR and LNER. A very prominent member of the Railway Club and an authority on early steam locomotives.

David Hunt: A former chairman of the Midland Railway Society and an outstanding modeller. Co-author with Peter Truman of *Midland Railway Portrait*. He has written many other books and articles on the Midland Railway.

W.H. Jameson: A professional engineer and a former prominent member of the Stephenson Locomotive Society, with a great interest in locomotive aesthetics.

Ronald Jarvis: A most distinguished steam locomotive designer Responsible for the rebuilding of the Bulleid pacifics, which I personally think should be called 'Jarvis pacifics'.

Eddie Johnson: A very well known author who has written a very considerable number of books on railways.

Kenneth Leech: He lived to be over one hundred years of age. He was the leading Great Northern Railway historian and in particular on Patrick Stirling's locomotives.

P. Marshall: The founder and editor of the *Model Engineer*, which began publication in 1898.

Derek Middleton: A current member of the Midland and Great Northern Joint Railway Circle, who has been writing about the locomotives of the M&GN Joint Railway since at least 1943.

John Minnis: A very well known modern writer with a great interest in William Stroudley and the London, Brighton and South Coast Railway.

O. S. Nock: 1905-1994. A world famous authority, and author on the steam locomotive, who had also travelled abroad extensively.

Vaughan Pendred: Editor of *The Engineer* for many years who was succeeded by his son Loughnan Pendred from 1905 to 1946. Loughnan's son Benjamin then took over the reigns

E.C. Poultney: A distinguished engineer who commenced his career with the Furness Railway at the beginning of the twentieth century.

Brian Radford: A former professional railwayman, and an authority on the Midland Railway.

Iain Rice: A well known modeller and writer of today.

Colonel H.C.B. Rogers: A very well known writer on steam locomotives. Author of books on Bulleid, Thompson, Peppercorn and André Chapelon.

Stephen Ross: Undoubtedly one of the finest locomotive modellers of today, with a great love of Victorian and Edwardian locomotives.

Charles Rous-Marten: 1844-1908. Undoubtedly one of the greatest Victorian / Edwardian railway writers.

John Rowland: Has been responsible for many informative letters to the railway press since the nineteen fifties.

Professor Jack Simmons: Died in 2000 in his late eighties. Considered by many to be the finest modern writer on railways. He considered the late Sir William Acworth to be the finest of all railway writers.

David Tee: 1928-2001. The leading Midland Railway locomotive historian of his time. A worthy successor to the late Paul C Dewhurst.

William Briggs Thompson: 1867-1962. A barrister and a man of independent means who travelled widely in the USA, Canada, Africa and Europe, and (because of his long life) probably saw more varied and attractive locomotives than any other person.

Howard Turner: A very well known lecturer and a true lover of the steam locomotive, particularly those of the Victorian and Edwardian era.

Robert Weatherburn: A most distinguished engineer who served the Midland Railway for very many years with a great knowledge of locomotive history, not only in this country but also abroad.

Peter Winding: An authority on the London Brighton & South Coast; London Chatham & Dover and the London & South Western railways.

I would like to express my most grateful thanks to my good friend John Musselwhite for typing the text of this address.